JOHANNES KEPLER

GREAT ASTRONOMERS

CW01551621

ROBERT STAWELL BALL

GIFT *Certificate*

TO:

_____

_____

_____

FROM:

_____

_____

DATE: _____

Would you like to buy a copy of
JOHANNES KEPLER?

PLEASE VISIT:
http://www.diamondbooks.ca

JOHANNES KEPLER

GREAT ASTRONOMERS

BY

ROBERT STAWELL BALL

D.Sc, LL.D., F.R.S.

(July 1, 1840 – November 25, 1913)

LOWNDEAN PROFESSOR OF ASTRONOMY AND GEOMETRY
IN THE UNIVERSITY OF CAMBRIDGE

www.diamondbooks.ca

TORONTO, CANADA – 2017

DIAMOND **BOOKS - CANADA**

Toronto, ON, CANADA
http://www.**diamond**books.ca

BIBLIOGRAPHIC INFORMATION

'JOHANNES KEPLER'
Is an extract or a chapter from the book
"GREAT ASTRONOMERS" by Robert Stawell Ball.

FIRST PUBLISHED IN: 1896.

PUBLISHED IN CANADA

Published in Canada by DIAMOND BOOKS - CANADA, an imprint of
DIAMOND PUBLISHERS - http://www.diamondpublishers.com

FIRST EDITION: MARCH, 2017.

PAPERBACK EDITION : ISBN: 978-1-988357-36-2

PRINTED IN CANADA

PREFACE
TO THE ORIGINAL EDITION

It has been my object in these pages to present the life of each astronomer in such detail as to enable the reader to realise in some degree the man's character and surroundings; and I have endeavoured to indicate as clearly as circumstances would permit the main features of the discoveries by which he has become known.

There are many types of astronomers—from the stargazer who merely watches the heavens, to the abstract mathematician who merely works at his desk; it has, consequently, been necessary in the case of some lives to adopt a very different treatment from that which seemed suitable for others.

ROBERT STAWELL BALL.

The Observatory, Cambridge.
October, 1895.

NOTE: The full-length of 'Preface' can be found in its original edition of the book "GREAT ASTRONOMERS".

- ROBERT STAWELL BALL

INTRODUCTION

Of all the natural sciences there is not one which offers such sublime objects to the attention of the inquirer as does the science of astronomy. From the earliest ages the study of the stars has exercised the same fascination as it possesses at the present day. Among the most primitive peoples, the movements of the sun, the moon, and the stars commanded attention from their supposed influence on human affairs. The practical utilities of astronomy were also obvious in primeval times. Maxims of extreme antiquity show how the avocations of the husbandman are to be guided by the movements of the heavenly bodies. The positions of the stars indicated the time to plough, and the time to sow. To the mariner who was seeking a way across the trackless ocean, the heavenly bodies .offered the only reliable marks by which his path could be guided. There was, accordingly, a stimulus both from intellectual curiosity and from practical necessity to follow the movements of the stars. Thus began a search for the causes of the ever-varying phenomena which the heavens display.

- ROBERT STAWELL BALL

Many of the earliest discoveries are indeed prehistoric. The great diurnal movement of the heavens, and the annual revolution of the sun, seem to have been known in times far more ancient than those to which any human monuments can be referred. The acuteness of the early observers enabled them to single out the more important of the wanderers which we now call planets. They saw that the star-like objects, Jupiter, Saturn, and Mars, with the more conspicuous Venus, constituted a class of bodies wholly distinct from the fixed stars among which their movements lay, and to which they bear such a superficial resemblance. But the penetration of the early astronomers went even further, for they recognized that Mercury also belongs to the same group, though this particular object is seen so rarely. It would seem that eclipses and other phenomena were observed at Babylon from a very remote period, while the most ancient records of celestial observations that we possess are to be found in the Chinese annals. The study of astronomy, in the sense in which we understand the word, may be said to have commenced under the reign of the Ptolemies at Alexandria. The most famous name in the science of this period is that of Hipparchus, who lived and worked at Rhodes about the year 160 B.C. It was his splendid investigations that first wrought the observed facts into a coherent branch of knowledge. He recognized the primary obligation which lies on the student of the heavens to compile as complete an inventory as possible of the objects which are there to be found. Hipparchus accordingly commenced by undertaking, on a small scale, a task exactly similar to that on which modern astronomers, with all available appliances of meridian circles, and photographic telescopes, are constantly engaged at the present day. He compiled a catalogue of the principal fixed stars, which is of special value to astronomers, as being the earliest work of its kind which has been handed down. He also studied the movements of the sun and the moon, and framed theories to account for the incessant changes which he saw in progress. He found a much more difficult problem in his attempt to interpret satisfactorily the complicated movements of the planets. With the view of constructing a theory which should give some coherent account of the subject, he made many observations of the places of these wandering stars. How great were the advances which Hipparchus accomplished may be appreciated if we reflect that, as a preliminary task to his

- ROBERT STAWELL BALL

more purely astronomical labours, he had to invent that branch of mathematical science by which alone the problems he proposed could be solved. It was for this purpose that he devised the indispensable method of calculation which we now know so well as trigonometry. Without the aid rendered by this beautiful art it would have been impossible for any n ally important advance in astronomical calculation to have been effected.

But the discovery which shows, beyond all others, that Hipparchus possessed one of the master-minds of all time was the detection of that remarkable celestial movement known as the precession of the equinoxes. The inquiry which conducted to this discovery involved a most profound investigation, especially when it is remembered that in the days of Hipparchus the means of observation of the heavenly bodies were only of the rudest description, and the available observations of earlier dates were extremely scanty. We can but look with astonishment on the genius of the man who, in spite of such difficulties, was able to detect such a phenomenon as the precession, and to exhibit its actual magnitude. I shall endeavour to explain the nature of this singular celestial movement, for it may be said to offer the first instance in the history of science in which we find that combination of accurate observation with skilful interpretation, of which, in the subsequent development of astronomy, we have so many splendid examples.

The word equinox implies the condition that the night is equal to the day. To a resident on the equator the night is no doubt equal to the day at all times in the year, but to one who lives on any other part of the earth, in either hemisphere, the night and the day are not generally equal. There is, however, one occasion in spring, and another in autumn, on which the day and the night are each twelve hours at all places on the earth. When the night and day are equal in spring, the point which the sun occupies on. the heavens is termed the vernal equinox. There is similarly another point in which the sun is situated at the time of the autumnal equinox. In any investigation of the celestial movements the positions of these two equinoxes on the heavens are of primary importance, and Hipparchus, with the instinct of genius, perceived their significance, and commenced to study them. It will be understood that we can always define the position of a point on the sky with reference to the surrounding stars. No doubt we do not see the stars near the sun when the sun is shining, but they are there nevertheless. The ingenuity of Hipparchus enabled him to determine the positions of each of the two equinoxes relatively to the stars which lie in its immediate vicinity. After examination of the celestial places of these points at different periods, he was led to the conclusion that each equinox was moving relatively to the stars, though that movement was so slow that twenty five thousand years would necessarily elapse before a complete circuit of the heavens was accomplished. Hipparchus traced out this phenomenon, and established it on an impregnable basis, so that all astronomers have ever since recognised the precession of the equinoxes) as one of the fundamental facts of astronomy. Not until nearly two thousand years after Hipparchus had made this splendid discovery was the explanation of its cause given by Newton.

From the days of Hipparchus down to the present hour the science of astronomy has steadily grown. One great observer after another has appeared from time to time, to reveal some new phenomenon with regard to the celestial bodies or their movements, while from time to time one commanding intellect after another has arisen to explain the true import of the facts of observations. The history of astronomy thus becomes inseparable from the history of the great men to whose labours its development is due. In the ensuing chapters we have endeavoured to sketch the lives and the work of the great philosophers, by whose labours the science of astronomy has been created.

We shall commence with Ptolemy, who, after the foundations of the science had been laid by Hipparchus, gave to astronomy the form in which it was taught throughout the Middle Ages. We shall next see the mighty revolution in our conceptions of the universe which are associated with the name of Copernicus. We then pass to those periods illumined by the genius of Galileo and Newton, and afterwards we shall trace the careers of other more recent discoverers, by whose industry and genius the boundaries of human knowledge have been so greatly extended. Our history will be brought down late enough to include some of the illustrious astronomers who laboured in the generation which has just passed away.

- ROBERT STAWELL BALL

TABLE OF CONTENTS

CHAPTERS

*An original Chapter from the book:
"GREAT ASTRONOMERS", by Robert Stawell Ball.

- ROBERT STAWELL BALL

CHAPTER

5

GREAT ASTRONOMERS

Johannes Kepler

(December 27, 1571 – November 15, 1630)

- ROBERT STAWELL BALL

Johannes Kepler
(December 27, 1571 – November 15, 1630)

CHAPTER
5

GREAT ASTRONOMERS:
JOHANNES KEPLER

WHILE the illustrious astronomer, Tycho Brahe, lay on his death-bed, he had an interview which must ever rank as one of the important incidents in the history of science. The life of Tycho had been passed, as we have seen, in the accumulation of vast stores of careful observations of the positions of the heavenly bodies. It was not given to him to deduce from his splendid work the results to which they were destined to lead. It was reserved for another astronomer to distil, so to speak, from the volumes in which Tycho's figures were recorded, the great truths of the universe which those figures contained. Tycho felt that his work required an interpreter, and he recognised in the genius of a young man with whom he was acquainted the agent by whom the world was to be taught some of the great truths of nature. To the bedside of the great Danish astronomer the youthful philosopher was summoned, and with his last breath Tycho besought of him to spare no labour in the performance of those calculations, by which alone the secrets of the move-ments of the heavens could be revealed. The solemn trust thus imposed was duly accepted, and the man who accepted it bore the immortal name of Kepler.

Kepler was born on the 27th December, 1571, at Weil, in the Duchy of Würtemberg. It would seem that the circumstances of his childbood must have been singularly-unhappy. His father, sprung from a well-connected family, was but a shiftless and idle adventurer; nor was the great astronomer much more fortunate in his other parent. His mother was an ignorant and ill-tempered woman; indeed, the ill-assorted union came to an abrupt end through the desertion of the wife by her husband when their eldest son Johannes, the hero of our present sketch, was eighteen years old. The childhood of this lad, destined for such fame, was still further embittered by the circumstance that when he was four years old he had a severe attack of small-pox. Not only was his eyesight permanently injured, but even his constitution appears to have been much weakened by this terrible malady.

It seems, however, that the bodily infirmities of young Johannes Kepler were the immediate cause of his attention being directed to the pursuit of knowledge. Had the boy been fitted like other boys for ordinary manual work, there can be hardly any doubt that to manual work his life must have been devoted. But, though his body was feeble, he soon gave indications of the possession of considerable mental power. It was accordingly thought that a suitable sphere for his talents might be found in the Church, which, in those days, was almost the only profession that afforded an opening for an intellectual career. We thus find that by the time Johannes Kepler was seventeen years old he had attained a sufficient standard of knowledge to entitle him to admission on the foundation of the University at Tübingen.

In the course of his studies at this institution he seems to have divided his attention equally between astronomy and divinity. It not unfrequently happens that when a man has attained considerable proficiency in two branches of knowledge he is not able to see very clearly in which of the two pursuits his true vocation lies. His friends and onlookers are often able to judge more wisely than he himself can do as to which of the two lines it would be better for him to pursue. This incapacity for perceiving the path in which greatness awaited him, existed in the case of Kepler. Personally, he inclined to enter the ministry, in which a promising career seemed open to him. He yielded, however, to friends, who evidently knew him better than he knew himself, and accepted, in 1594, the important professorship of astronomy which had been offered to him in the University of Grätz.

It is difficult for us in these modern days to realise the somewhat extraordinary duties which were expected from an astronomical professor in the sixteenth century. He was, of course, required to employ his knowledge of the heavens in the prediction of eclipses, and of the movements of the heavenly bodies generally. This seems reasonable enough; but what we are not prepared to accept is the obligation which lay on the astronomers to predict the fates of nations and the destinies of individuals.

It must be remembered that it was the almost universal belief in those days, that all the celestial spheres revolved in some mysterious fashion around the earth, which appeared by far the most important body in the universe. It was imagined that the sun, the moon, and the stars indicated, in the vicissitudes of their movements, the careers of nations and of individuals. Such being the generally accepted notion, it seemed to follow that a professor who was charged with the duty of expounding the movements of the heavenly bodies must necessarily be looked to for the purpose of deciphering the celestial decrees regarding the fate of man which the heavenly luminaries were designed to announce.

Kepler threw himself with characteristic ardour into even this fantastic phase of the labours of the astronomical professor; he diligently studied the rules of astrology, which the fancies of antiquity had compiled. Believing sincerely as he did in the connection between the aspect of the stars and the state of human affairs, he even thought that he perceived, in the events of his own life, a corroboration of the doctrine which affirmed the influence of the planets upon the fate of individuals.

But quite independently of astrology there seem to have been many other delusions current among the philosophers of Kepler's time. It is now almost incomprehensible how the ablest men of a few centuries ago should have entertained such preposterous notions, as they did, with respect to the sys-tem of the universe. As an instance of what is here referred to, we may cite the extraordinary notion which, under the designation of a discovery, first brought Kepler into fame. Geometers had long known that there were five, but no more than five, regular solid figures. There is, for instance, the cube with six sides, which is, of course, the most familiar of these solids.

Besides the cube there are other figures of

four, eight, twelve, and twenty sides respectively. It also happened that there were five planets, but no more than five, known to the ancients, namely. Mercury, Venus, Mars, Jupiter, and Saturn. To Kepler's lively imaginations this coincidence suggested the idea that the five regular solids corresponded to the five planets, and a number of fancied numerical relations were adduced on the subject.

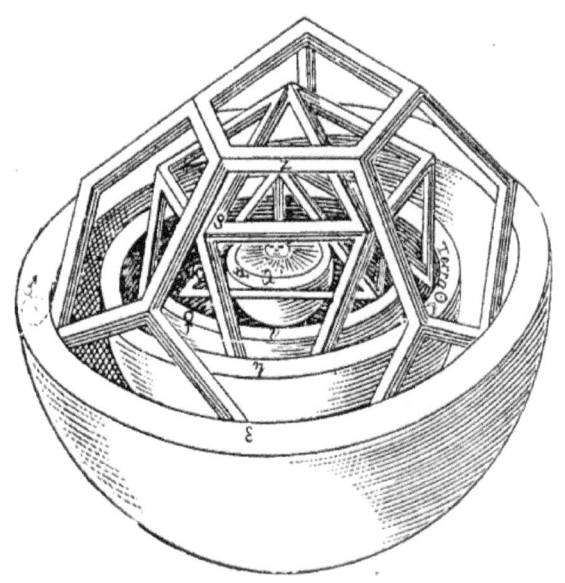

Kepler's system of regular solids.

The absurdity of this doctrine is obvious enough, especially when we observe that, as is now well known, there are two large planets, and a host of small planets, over and above the magical number of the regular solids. In Kepler's time, however, this doctrine was so far from being regarded as absurd, that its announcement was hailed as a great intellectual triumph. Kepler was at once regarded with favour. It seems, indeed, to have been the circumstance which brought him into correspondence with Tycho Brahe. By its means also he became known to Galileo.

The career of a scientific professor in those early days appears generally to have been marked by rather more striking vicissitudes than usually befall a professor in a modern university. Kepler was a Protestant, and as such he had been appointed to his professorship at Grätz. A change, however, having taken place in the religious belief entertained by the ruling powers of the University, the Protestant professors were expelled. It seems that special influence having been exerted in Kepler's case on account of his exceptional eminence, he was recalled to Grätz, and reinstated in the tenure of his chair. But his pupils had vanished, so that the great astronomer was glad to accept a post offered him by Tycho Brahe in the observatory which the latter had recently established near Prague.

On Tycho's death, which occurred soon after, an opening presented itself which gave Kepler the opportunity his genius demanded. He was appointed to succeed Tycho in the position of imperial mathematician. But a far more important point, both for Kepler and for science, was that to him was confided the use of Tycho's observations. It was, indeed, by the discussion of Tycho's results that Kepler was enabled to make the discoveries which form such an important part of astronomical history.

Kepler must also be remembered as one of the first great astronomers who ever had the privilege of viewing celestial bodies through a telescope. It was in 1610 that he first held in his hands one of those little instruments which had been so recently applied to the heavens by Galileo. It should, however, be borne in mind that the epoch-making achievements of Kepler did not arise from any telescopic observations that he made, or, indeed, that any one else made. They were all elaborately deduced from Tycho's measurements of the positions of the planets, obtained with his great instruments, which were unprovided with telescopic assistance.

To realise the tremendous advance which science received from Kepler's great work, it is to be understood that all the astronomers who laboured before him at the difficult subject of the celestial motions, took it for granted that the planets must revolve in circles. If it did not appear that a planet moved in a fixed circle, then the ready answer was provided by Ptolemy's theory that the circle in which the planet did move was itself in motion, so that its centre described another circle.

When Kepler had before him that wonderful series of observations of the planet Mars, which had been accumulated by the extraordinary skill of Tycho, he proved, after much labour, that the movements of the planet refused to be represented in a circular form. Nor would it do to suppose that Mars revolved in one circle, the centre of which revolved in another circle. On no such supposition could the movements of the planets be made to tally with those which Tycho had actually observed. This led to the astonishing discovery of the true form of a planet's orbit. For the first time in the history of astronomy the principle was laid down that the movement of a planet could not be represented by a circle, nor even by combinations of circles, but that it could be represented by an elliptic path. In this path the sun is situated at one of those two points in the ellipse which are known as its foci.

Very simple apparatus is needed for the drawing of one of those ellipses which Kepler has shown to possess such astonishing astronomical significance. Two pins are stuck through a sheet of paper on a board, the point of a pencil is inserted in a loop of string which passes over the pins, and as the pencil is moved round in such a way as to keep the string stretched, that beautiful curve known as the ellipse is delineated, while the positions of the pins indicate the two foci of the curve. If the length of the loop of string is unchanged then the nearer the pins are together, the greater will be the resemblance between the ellipse and the circle, whereas the more the pins are separated the more elongated does the ellipse become. The orbit of a great planet is, in general, one of those ellipses which approaches a nearly circular form. It fortunately happens, however, that the orbit of Mars makes a wider departure from the circular form than any of the other important planets. It is, doubtless, to this circumstance that we must attribute the astonishing success of Kepler in detecting the true shape of a planetary orbit. Tycho's observations would not have been sufficiently accurate to have exhibited the elliptic nature of a planetary orbit which, like that of Venus, differed very little from a circle.

The more we ponder on this memorable achievement the more striking will it appear. It must be remembered that in these days we know of the physical necessity which requires that a planet shall revolve in an ellipse and not in any other curve. But Kepler had no such knowledge. Even to the last hour of his life he remained in ignorance of the existence of any natural cause which ordained that planets should follow those particular curves which geometers know so well.

Kepler's assignment of the ellipse as the true form of the planetary orbit is to be regarded as a brilliant guess, the truth of which Tycho's observations enabled him to verify. Kepler also succeeded in pointing out the law according to which the velocity of a planet at different points of its path could be accurately specified. Here, again, we have to admire the sagacity with which this marvellously acute astronomer guessed the deep truth of nature. In this case also he was quite unprovided with any reason for expecting from physical principles that such a law as he discovered must be obeyed. It is quite true that Kepler had some slight knowledge of the existence of what we now know as gravitation. He had even enunciated the remarkable doctrine that the ebb and flow of the tide must be attributed to the attraction of the moon on the waters of the earth. He does not, however, appear to have had any anticipation of those wonderful discoveries which Newton was destined to make a little later, in which he demonstrated that the laws detected by Kepler's marvellous acumen were necessary consequences of the principle of universal gravitation.

To appreciate the relations of Kepler and Tycho it is necessary to note the very different way in which these illustrious astronomers viewed the system of the heavens. It should be observed that Copernicus had already expounded the true system, which located the sun at the centre of the planetary system. But in the days of Tycho Brahe this doctrine had not as yet commanded universal assent. In fact, the great observer himself did not accept the new views of Copernicus. It appeared to Tycho that the earth not only appeared to be the centre of things celestial, but that it actually was the centre. It is, indeed, not a little remarkable that a student of the heavens so accurate as Tycho should have deliberately rejected the Copernican doctrine in favour of the system which now seems so preposterous.

Throughout his great career, Tycho steadily observed the places of the sun, the moon, and the planets, and as steadily maintained that all those bodies revolved around the earth fixed in the centre. Kepler, however, had the advantage of belonging to the new school.

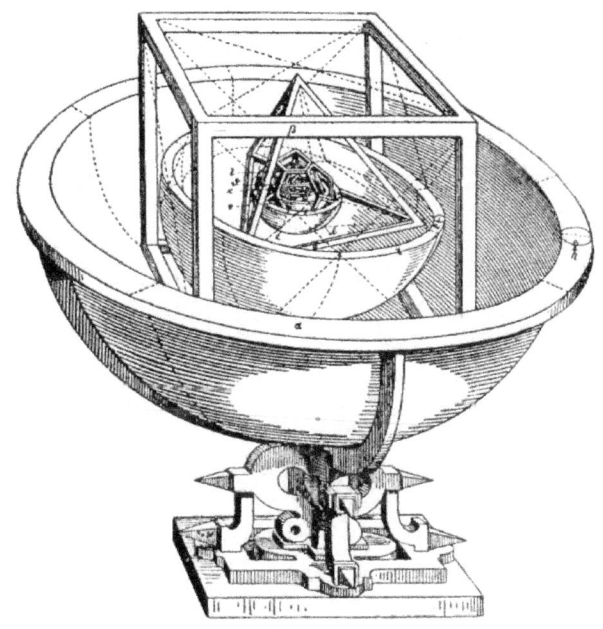

Symbolical representation of the planetary system.

He utilised the observations of Tycho in developing the great Copernican theory whose teaching Tycho stoutly resisted.

Perhaps a chapter in modern science may illustrate the intellectual relation of these great men. The revolution produced by Copernicus in the doctrine of the heavens has often been likened to the revolution which the Darwinian theory produced in the views held by biologists as to life on this earth. The Darwinian theory did not at first command universal assent even among those naturalists whose lives had been devoted with the greatest success to the study of organisms. Take, for instance, that great naturalist, Professor Owen,5 by whose labours vast extension has been given to our knowledge of the fossil animals which dwelt on the earth in past ages. Now, though Owen's researches were intimately connected with the great labours of Darwin, and afforded the latter material for his epoch-making generalization, yet Owen deliberately refused to accept the new doctrines. Like Tycho, he kept on rigidly accumulating his facts under the influence of a set of ideas as to the origin of living forms which are now universally admitted to be erroneous. If, therefore, we liken Darwin to Copernicus, and Owen to Tycho, we may liken the biologists of the present day to Kepler, who interpreted the results of accurate observation upon sound theoretical principles.

In reading the works of Kepler in the light of our modern knowledge we are often struck by the extent to which his perception of the sublimest truths in nature was associated with the most extravagant errors and absurdities. But, of course, it must be remembered that he wrote in an age in which even the rudiments of science, as we now understand it, were almost entirely unknown.

- ROBERT STAWELL BALL

It may well be doubted whether any joy experienced by mortals is more genuine than that which rewards the successful searcher after natural truths. Every science-worker, be his efforts ever so humble, will be able to sympathise with, the enthusiastic delight of Kepler when at last, after years of toil, the glorious light broke forth, and that which, he considered to be the greatest of his astonishing laws first dawned upon him. Kepler rightly judged that the number of days which a planet required to perform its voyage round the sun must be connected in some manner with the distance from the planet to the sun; that is to say, with the radius of the planet's orbit, inasmuch as we may for our present object regard the planet's orbit as circular.

Here, again, in his search for the unknown law, Kepler had no accurate dynamical principles to guide his steps. Of course, we now know not only what the connection between the planet's distance and the planet's periodic time actually is, but we also know that it is a necessary consequence of the law of universal gravitation. Kepler, it is true, was not without certain surmises on the subject, but they were of the most fanciful description. His notions of the planets, accurate as they were in certain important respects, were mixed up with vague ideas as to the properties of metals and the geometrical relations of the regular solids. Above all, his reasoning was penetrated by the supposed astrological influences of the stars and their significant relation to human fate. Under the influence of such a farrago of notions, Kepler resolved to make all sorts of trials in his search for the connection between the distance of a planet from the sun and the time in which the revolution of that planet was accomplished.

It was quite easily demonstrated that the greater the distance of the planet from the sun the longer was the time required for its journey. It might have been thought that the time would be directly proportional to the distance. It was, however, easy to show that this supposition did not agree with the fact. Finding that this simple relation would not do, Kepler undertook a vast series of calculations to find out the true method of expressing the connection. At last, after many vain attempts, he found, to his indescribable joy, that the square of the time in which a planet revolves around the sun was proportional to the cube of the average distance of the planet from that body.

The extraordinary way in which Kepler's views on celestial matters were associated with the wildest speculations, is well illustrated in the work in which he propounded his splendid discovery just referred to. The announcement of the law connecting the distances of the planets from the sun with their periodic times, was then mixed up with a preposterous conception about the properties of the different planets. They were supposed to be associated with some profound music of the spheres inaudible to human ears, and performed only for the benefit of that being whose soul formed the animating spirit of the sun.

Kepler was also the first astronomer who ever ventured to predict the occurrence of that remarkable phenomenon, the transit of a planet in front of the sun's disc. He published, in 1629, a notice to the curious in things celestial, in which he announced that both of the planets. Mercury and Venus, were to make a transit across the sun on specified days in the winter of 1631. The transit of Mercury was duly observed by Gassendi,7 and the transit of Venus also took place, though, as we now know, the circumstances were such that it was not possible for the phenomenon to be witnessed by any European astronomer.

In addition to Kepler's discoveries already mentioned, with which his name will be forever associated, his claim on the gratitude of astronomers chiefly depends on the publication of his famous Rudolphine tables. In this remarkable work means are provided for finding the places of the planets with far greater accuracy than had previously been attainable.

Kepler, it must be always remembered, was not an astronomical observer. It was his function to deal with the observations made by Tycho, and, from close study and comparison of the results, to work out the movements of the heavenly bodies. It was, in fact, Tycho who provided as it were the raw material, while it was the genius of Kepler which wrought that material into a beautiful and serviceable form. For more than a century the Rudolphine tables were regarded as a standard astronomical work. In these days we are accustomed to find the movements of the heavenly bodies set forth with all desirable exactitude in the *Nautical Almanack*, and the similar publication issued by foreign Grovernments. Let it be remembered that it was Kepler who first imparted the proper impulse in this direction.

When Kepler was twenty-six he married an heiress from Styria, who, though only twenty-three years old, had already had some experience in matrimony. Her first husband had died; and it was after her second husband had divorced her that she received the addresses of Kepler. It will not be surprising to hear that his domestic affairs do not appear to have been particularly happy, and his wife died in 1611. Two years later, undeterred by the want of success in his first venture, he sought a second partner, and he evidently determined not to make a mistake this time. Indeed, the methodical manner in which he made his choice of the lady to whom he should propose has been duly set forth by him and preserved for our edification. With some self-assurance he asserts that there were no fewer than eleven spinsters desirous of sharing his joys and sorrows. He has carefully estimated and recorded the merits and demerits of each of these would-be brides. The result of his deliberations was that he awarded himself to an orphan girl, destitute even of a portion.

**The Commemoration of the
Rudolphine Tables.**

- ROBERT STAWELL BALL

Success attended his choice, and his second marriage seems to have proved a much more suitable union than his first. He had five children by the first wife and seven by the second.

The years of Kepler's middle life were sorely distracted by a trouble which, though not uncommon in those days, is one which we find it difficult to realise at the present time. His mother, Catherine Kepler, had attained undesirable notoriety by the suspicion that she was guilty of witchcraft. Years were spent in legal investigations, and it was only after unceasing exertions on the part of the astronomer for upwards of a twelvemonth that he was finally able to procure her acquittal and release from prison.

It is interesting for us to note that at one time there was a proposal that Kepler should forsake his native country and adopt England as a home. It arose in this wise. The great man was distressed throughout the greater part of his life by pecuniary anxieties. Finding him in a strait of this description, the English ambassador in Venice, Sir Henry Wotton, in the year 1620, besought Kepler to come over to England, where he assured him that he would obtain a favourable reception, and where, he was able to add, Kepler's great scientific work was already highly esteemed. But his efforts were unavailing; Kepler would not leave his own country. He was then forty-nine years of age, and doubtless a home in a foreign land, where people spoke a strange tongue, had not sufficient attraction for him, even when accompanied with the substantial inducements which the ambassador was able to offer. Had Kepler accepted this invitation, he would, in transferring his home to England, have anticipated the similar change which took place in the career of another great astronomer two centuries later. It will be remembered that Herschel, in his younger days, did transfer himself to England, and thus gave to

England the imperishable fame of association with his triumphs.

The publication of the Rudolphine tables of the celestial movements entailed much expense. A considerable part of this was defrayed by the Government at Venice, but the balance occasioned no little trouble and anxiety to Kepler. No doubt the authorities of those days were even less willing to spend money on scientific matters than are the Governments of more recent times. For several years the imperial Treasury was importuned to relieve him from his anxieties. The effects of so much worry, and of the long journeys which were involved, at last broke down Kepler's health completely. As we have already mentioned, he had never been strong from infancy, and he finally succumbed to a fever in November, 1630, at the age of fifty-nine. He was interred at St. Peter's Church, at Ratisbon.

Though Kepler had not those personal characteristics which have made his great predecessor, Tycho Brahe, such a romantic figure, yet a picturesque element in Kepler's character is not wanting. It was, however, of an intellectual kind. His imagination, as well as his reasoning faculties, always worked together. He was incessantly prompted by the most extraordinary speculations. The great majority of them were in a high degree wild and chimerical, but every now and then one of his fancies struck right to the heart of nature, and an immortal truth was brought to light.

I remember visiting the observatory of one of our greatest modern astronomers, and in a large desk he showed me a multitude of photographs which he had attempted but which had not been successful, and then he showed me the few and rare pictures which had succeeded, and by which important truths had been revealed. With a felicity of expression which I have often since thought of, he alluded to the contents of the desk as the "chips." They were useless, but they were necessary incidents in the truly successful work. So it is in all great and good work. Even the most skilful man of science pursues many a wrong scent. Time after time he goes off on some track that plays him false. The greater the man's genius and intellectual resource, the more numerous will be the ventures which he makes, and the great majority of those ventures are certain to be fruitless. They are, in fact, the "chips." In Kepler's case the chips were numerous enough. They were of the most extraordinary variety and structure. But every now and then a sublime discovery was made of such a character as to make us regard even the most fantastic of Kepler's chips with the greatest veneration and respect.

COMPLIMENTARY MATERIAL

★

A BRIEF BIOGRAPHICAL SKETCH

JOHANNES KEPLER

★

- ROBERT STAWELL BALL

Johannes Kepler
(December 27, 1571 – November 15, 1630)

★

A BRIEF BIOGRAPHICAL SKETCH

JOHANNES KEPLER

★

Johannes Kepler (December 27, 1571 – November 15, 1630) was a German mathematician, astronomer, and astrologer. A key figure in the 17th-century scientific revolution, he is best known for his laws of planetary motion, based on his works Astronomia nova, Harmonices Mundi, and Epitome of Copernican Astronomy. These works also provided one of the foundations for Isaac Newton's theory of universal gravitation.

Kepler was a mathematics teacher at a seminary school in Graz, where he became an associate of Prince Hans Ulrich von Eggenberg. Later he became an assistant to the astronomer Tycho Brahe in Prague, and eventually he was the imperial mathematician to Emperor Rudolf II and his two successors Matthias and Ferdinand II. He was also a mathematics teacher in Linz, and an adviser to General Wallenstein. Additionally, he did fundamental work in the field of optics, invented an improved version of the refracting telescope (the Keplerian telescope), and was mentioned in the telescopic discoveries of his contemporary Galileo Galilei.

Kepler lived in an era when there was no clear distinction between astronomy and astrology, but there was a strong division between astronomy (a branch of mathematics within the liberal arts) and physics (a branch of natural philosophy). Kepler also incorporated religious arguments and reasoning into his work, motivated by the religious conviction and belief that God had created the world according to an intelligible plan that is accessible through the natural light of reason. Kepler described his new astronomy as "celestial physics", as "an excursion into Aristotle's Metaphysics", and as "a supplement to Aristotle's On the

Heavens", transforming the ancient tradition of physical cosmology by treating astronomy as part of a universal mathematical physics.

Early Years

Kepler was born on December 27, the feast day of St John the Evangelist, 1571, in the Free Imperial City of Weil der Stadt (now part of the Stuttgart Region in the German state of Baden-Württemberg, 30 km west of Stuttgart's center). His grandfather, Sebald Kepler, had been Lord Mayor of the city. By the time Johannes was born, he had two brothers and one sister and the Kepler family fortune was in decline. His father, Heinrich Kepler, earned a precarious living as a mercenary, and he left the family when Johannes was five years old. He was believed to have died in the Eighty Years' War in the Netherlands. His mother Katharina Guldenmann, an innkeeper's daughter, was a healer and herbalist. Born prematurely, Johannes claimed to have been weak and sickly as a child. Nevertheless, he often impressed travelers at his grandfather's inn with his phenomenal mathematical faculty.

Kepler's birthplace, in **Weil der Stadt**

- ROBERT STAWELL BALL

He was introduced to astronomy at an early age, and developed a love for it that would span his entire life. At age six, he observed the Great Comet of 1577, writing that he "was taken by [his] mother to a high place to look at it." In 1580, at age nine, he observed another astronomical event, a lunar eclipse, recording that he remembered being "called outdoors" to see it and that the moon "appeared quite red". However, childhood smallpox left him with weak vision and crippled hands, limiting his ability in the observational aspects of astronomy.

In 1589, after moving through grammar school, Latin school, and seminary at Maulbronn, Kepler attended Tübinger Stift at the University of Tübingen. There, he studied philosophy under Vitus Müller and theology under Jacob Heerbrand (a student of Philipp Melanchthon at Wittenberg), who also taught Michael Maestlin while he was a student, until he became Chancellor at Tübingen in 1590. He proved himself to be a superb mathematician and earned a reputation as a skilful astrologer, casting horoscopes for fellow students. Under the instruction of Michael Maestlin, Tübingen's professor of mathematics from 1583 to 1631, he learned both the Ptolemaic system and the Copernican system of planetary motion. He became a Copernican at that time. In a student disputation, he defended heliocentrism from both a theoretical and theological perspective, maintaining that the Sun was the principal source of motive power in the universe. Despite his desire to become a minister, near the end of his studies, Kepler was recommended for a position as teacher of mathematics and astronomy at the Protestant school in Graz. He accepted the position in April 1594, at the age of 23.

Graz (1594–1600)

Mysterium Cosmographicum

Kepler's first major astronomical work, Mysterium Cosmographicum (The Cosmographic Mystery) [1596], was the first published defense of the Copernican system. Kepler claimed to have had an epiphany on July 19, 1595, while teaching in Graz, demonstrating the periodic conjunction of Saturn and Jupiter in the zodiac: he realized that regular polygons bound one inscribed and one circumscribed circle at definite ratios, which, he reasoned, might be the geometrical basis of the universe. After failing to find a unique arrangement of polygons that fit known astronomical observations (even with extra planets added to the system), Kepler began experimenting with 3-dimensional polyhedra. He found that each of the five Platonic solids could be inscribed and circumscribed by spherical orbs; nesting these solids, each encased in a sphere, within one another would produce six layers, corresponding to the six known planets—Mercury, Venus, Earth, Mars, Jupiter, and Saturn. By ordering the solids selectively—octahedron, icosahedron, dodecahedron, tetrahedron, cube—Kepler found that the spheres could be placed at intervals corresponding to the relative sizes of each planet's path, assuming the planets circle the Sun. Kepler also found a formula relating the size of each planet's orb to the length of its orbital period: from inner to outer planets, the ratio of increase in orbital period is twice the difference in orb radius. However, Kepler later rejected this formula, because it was not precise enough.

As he indicated in the title, Kepler thought he had revealed God's geometrical plan for the universe. Much of Kepler's enthusiasm for the Copernican system stemmed from his theological convictions about the connection between the physical and the spiritual; the universe itself was an image of God, with the Sun corresponding to the Father, the stellar sphere to the Son, and the intervening space between to the Holy

Spirit. His first manuscript of Mysterium contained an extensive chapter reconciling heliocentrism with biblical passages that seemed to support geocentrism.

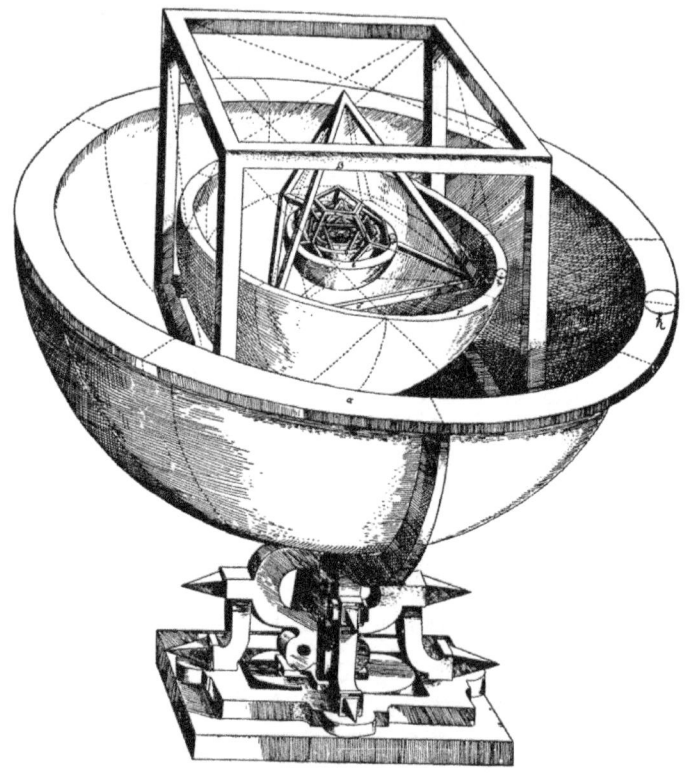

Kepler's Platonic solid model of the solar system, from Mysterium Cosmographicum (1596)

With the support of his mentor Michael Maestlin, Kepler received permission from the Tübingen university senate to publish his manuscript, pending removal of the Bible exegesis and the addition of a simpler, more understandable description of the Copernican system as well as Kepler's new ideas. Mysterium was published late in 1596, and Kepler received his copies and began sending them to prominent astronomers and patrons early in 1597; it was not widely read, but it established Kepler's reputation as a highly skilled astronomer. The effusive dedication, to powerful patrons as well as to the men who controlled his position in Graz, also provided a crucial doorway into the patronage system.

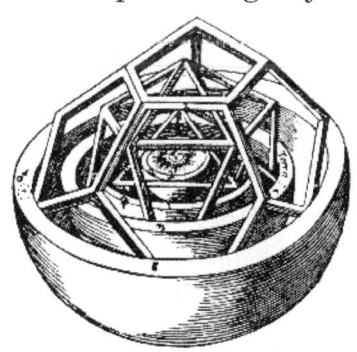

Close-up of an inner section of Kepler's model

Though the details would be modified in light of his later work, Kepler never relinquished the Platonist polyhedral-spherist cosmology of Mysterium Cosmographicum. His subsequent main astronomical works were in some sense only further developments of it, concerned with finding more precise inner and outer dimensions for the spheres by calculating the eccentricities of the planetary orbits within it. In 1621, Kepler published an expanded second edition of Mysterium, half as long again as the first, detailing in footnotes the corrections and improvements he had achieved in the 25 years since its first publication.

In terms of the impact of Mysterium, it can be seen as an important first step in modernizing the theory proposed by Nicolaus Copernicus in his "De Revolutionibus orbium coelestium". Whilst Copernicus sought to advance a heliocentric system in this book, he resorted to Ptolemaic devices (viz., epicycles and eccentric circles) in order to explain the change in planets' orbital speed, and also continued to use as a point of reference the center of the earth's orbit rather than that of the sun "as an aid to calculation and in order not to confuse the reader by diverging too much from Ptolemy." Modern astronomy owes much to "Mysterium Cosmographicum", despite flaws in its main thesis, "since it represents the first step in cleansing the Copernican system of the remnants of the Ptolemaic theory still clinging to it."

Marriage to Barbara Müller

In December 1595, Kepler was introduced to Barbara Müller, a 23-year-old widow (twice over) with a young daughter, Regina Lorenz, and he began courting her. Müller, heiress to the estates of her late husbands, was also the daughter of a successful mill owner. Her father Jobst initially opposed a marriage despite Kepler's nobility; though he had inherited his grandfather's nobility, Kepler's poverty made him an unacceptable match. Jobst relented after Kepler completed work on Mysterium, but the engagement nearly fell apart while Kepler was away tending to the details of publication. However, Protestant officials—who had helped set up the match—pressured the Müllers to honor their agreement. Barbara and Johannes were married on April 27, 1597.

Portraits of Kepler and his wife in oval medallions

In the first years of their marriage, the Keplers had two children (Heinrich and Susanna), both of whom died in infancy. In 1602, they had a daughter (Susanna); in 1604, a son (Friedrich); and in 1607, another son (Ludwig).

Other Research

Following the publication of Mysterium and with the blessing of the Graz school inspectors, Kepler began an ambitious program to extend and elaborate his work. He planned four additional books: one on the stationary aspects of the universe (the Sun and the fixed stars); one on the planets and their motions; one on the physical nature of planets and the formation of geographical features (focused especially on Earth); and one on the effects of the heavens on the Earth, to include atmospheric optics, meteorology, and astrology.

He also sought the opinions of many of the astronomers to whom he had sent Mysterium, among them Reimarus Ursus (Nicolaus Reimers Bär)—the imperial mathematician to Rudolph II and a bitter rival of Tycho Brahe. Ursus did not reply directly, but republished Kepler's flattering letter to pursue his priority dispute over (what is now called) the Tychonic system with Tycho. Despite this black mark, Tycho also began corresponding with Kepler, starting with a harsh but legitimate critique of Kepler's system; among a host of objections, Tycho took issue with the use of inaccurate numerical data taken from Copernicus. Through their letters, Tycho and Kepler discussed a broad range of astronomical problems, dwelling on lunar phenomena and Copernican theory (particularly its theological viability). But without the significantly more accurate data of Tycho's observatory, Kepler had no way to address many of these issues.

House of Kepler and Barbara Müller
in Gössendorf, near Graz (1597–1599)

Instead, he turned his attention to chronology and "harmony," the numerological relationships among music, mathematics and the physical world, and their astrological consequences. By assuming the Earth to possess a soul (a property he would later invoke to explain how the sun causes the motion of planets), he established a speculative system connecting astrological aspects and astronomical distances to weather and other earthly phenomena. By 1599, however, he again felt his work limited by the inaccuracy of available data—just as growing religious tension was also threatening his continued employment in Graz. In December of that year, Tycho invited Kepler to visit him in Prague; on January 1, 1600 (before he even received the invitation), Kepler set off in the hopes that Tycho's patronage could solve his philosophical problems as well as his social and financial ones.

- ROBERT STAWELL BALL

Prague (1600–1612)

Work for Tycho Brahe

On February 4, 1600, Kepler met Tycho Brahe and his assistants Franz Tengnagel and Longomontanus at Benátky nad Jizerou (35 km from Prague), the site where Tycho's new observatory was being constructed. Over the next two months he stayed as a guest, analyzing some of Tycho's observations of Mars; Tycho guarded his data closely, but was impressed by Kepler's theoretical ideas and soon allowed him more access. Kepler planned to test his theory from Mysterium Cosmographicum based on the Mars data, but he estimated that the work would take up to two years (since he was not allowed to simply copy the data for his own use). With the help of Johannes Jessenius, Kepler attempted to negotiate a more formal employment arrangement with Tycho, but negotiations broke down in an angry argument and Kepler left for Prague on April 6. Kepler and Tycho soon reconciled and eventually reached an agreement on salary and living arrangements, and in June, Kepler returned home to Graz to collect his family.

Political and religious difficulties in Graz dashed his hopes of returning immediately to Brahe; in hopes of continuing his astronomical studies, Kepler sought an appointment as mathematician to Archduke Ferdinand. To that end, Kepler composed an essay—dedicated to Ferdinand—in which he proposed a force-based theory of lunar motion: "In Terra inest virtus, quae Lunam ciet" ("There is a force in the earth which causes the moon to move"). Though the essay did not earn him a place in Ferdinand's court, it did detail a new method for measuring lunar eclipses, which he applied during the July 10 eclipse in Graz. These observations formed the basis of his explorations of the laws of optics that would culminate in Astronomiae Pars Optica.

Tycho Brahe

On August 2, 1600, after refusing to convert to Catholicism, Kepler and his family were banished from Graz. Several months later, Kepler returned, now with the rest of his household, to Prague. Through most of 1601, he was supported directly by Tycho, who assigned him to analyzing planetary observations and writing a tract against Tycho's (by then deceased) rival, Ursus. In September, Tycho secured him a commission as a collaborator on the new project he had proposed to the emperor: the Rudolphine Tables that should replace the Prutenic Tables of Erasmus Reinhold. Two days after Tycho's unexpected death on October 24, 1601, Kepler was appointed his successor as imperial mathematician with the responsibility to complete his unfinished work. The next 11 years as imperial mathematician would be the most productive of his life.

Advisor to Emperor Rudolph II

Kepler's primary obligation as imperial mathematician was to provide astrological advice to the emperor. Though Kepler took a dim view of the attempts of contemporary astrologers to precisely predict the future or divine specific events, he had been casting well-received detailed horoscopes for friends, family, and patrons since his time as a student in Tübingen. In addition to horoscopes for allies and foreign leaders, the emperor sought Kepler's advice in times of political trouble. Rudolph was actively interested in the work of many of his court scholars (including numerous alchemists) and kept up with Kepler's work in physical astronomy as well.

Officially, the only acceptable religious doctrines in Prague were Catholic and Utraquist, but Kepler's position in the imperial court allowed him to practice his Lutheran faith unhindered. The emperor nominally provided an ample income for his family, but the difficulties of the over-extended imperial treasury meant that actually getting hold of enough money to meet financial obligations was a continual struggle. Partly because of financial troubles, his life at home with Barbara was unpleasant, marred with bickering and bouts of sickness. Court life, however, brought Kepler into contact with other prominent scholars (Johannes Matthäus Wackher von Wackhenfels, Jost Bürgi, David Fabricius, Martin Bachazek, and

Johannes Brengger, among others) and astronomical work proceeded rapidly.

Astronomiae Pars Optica

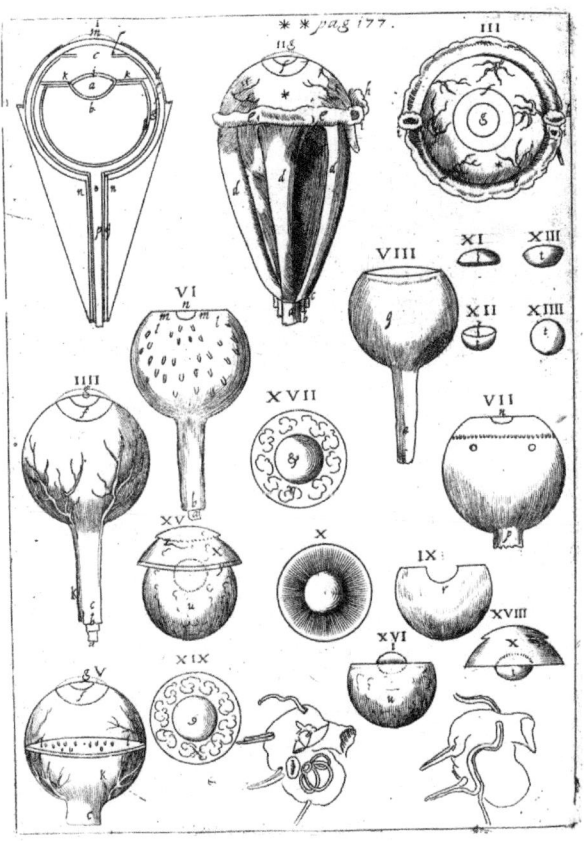

As he slowly continued analyzing Tycho's Mars observations—now available to him in their entirety—and began the slow process of tabulating the Rudolphine Tables, Kepler also picked up the investigation of the laws of optics from his lunar essay of 1600. Both lunar and solar eclipses presented unexplained phenomena, such as unexpected shadow sizes, the red color of a total lunar eclipse, and the reportedly unusual light surrounding a total solar eclipse. Related issues of atmospheric refraction applied to all astronomical observations. Through most of 1603, Kepler paused his other work to focus on optical theory; the resulting manuscript, presented to the emperor on January 1, 1604, was published as Astronomiae Pars Optica (The Optical Part of Astronomy). In it, Kepler described the inverse-square law governing the intensity of light, reflection by flat and curved mirrors, and principles of pinhole cameras, as well as the astronomical implications of optics such as parallax and the apparent sizes of heavenly bodies. He also extended his study of optics to the human eye, and is generally considered by neuroscientists to be the first to recognize that images are projected inverted and reversed by the eye's lens onto the retina. The solution to this dilemma was not of particular importance to Kepler as he did not see it as pertaining to optics, although he did suggest that the image was later corrected "in the hollows of the brain" due to the "activity of the Soul." Today, Astronomiae Pars Optica is generally recognized as the foundation of modern optics (though the law of refraction is conspicuously absent). With respect to the beginnings of projective geometry, Kepler introduced the idea of continuous change of a mathematical entity in this work. He argued that if a focus of a conic section were allowed to move along the line joining the foci, the geometric form would morph or degenerate, one into another. In this way, an ellipse becomes a parabola when a focus moves toward infinity, and when two foci of an ellipse merge into one another, a circle is formed. As the foci of a hyperbola merge into one another, the hyperbola becomes a pair of straight lines. He also assumed that if a straight line is extended to infinity it will meet itself at a single point at infinity, thus having the properties of a large circle.

The Supernova of 1604

In October 1604, a bright new evening star (SN 1604) appeared, but Kepler did not believe the rumors until he saw it himself. Kepler began systematically observing the nova. Astrologically, the end of 1603 marked the beginning of a fiery trigon, the start of the about 800-year cycle of great conjunctions; astrologers associated the two previous such periods with the rise of Charlemagne (c. 800 years earlier) and the birth of Christ (c. 1600 years earlier), and thus expected events of great portent, especially regarding the emperor. It was in this context, as the imperial mathematician and astrologer to the emperor, that Kepler described the new star two years later in his De Stella Nova. In it, Kepler addressed the star's astronomical properties while taking a skeptical approach to the many astrological interpretations then circulating. He noted its fading luminosity, speculated about its origin, and used the lack of observed parallax to argue that it was in the sphere of fixed stars, further undermining the doctrine of the immutability of the heavens (the idea accepted since Aristotle that the celestial spheres were perfect and unchanging). The birth of a new star implied the variability of the heavens. In an appendix, Kepler also discussed the recent chronology work of the Polish historian Laurentius Suslyga; he calculated that, if Suslyga was correct that accepted timelines were four years behind, then the Star of Bethlehem—analogous to the present new star—would have coincided with the first great conjunction of the earlier 800-year cycle.

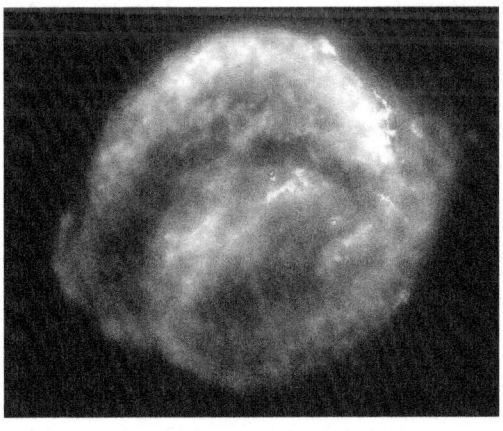

- ROBERT STAWELL BALL

Astronomia Nova

The extended line of research that culminated in Astronomia nova (A New Astronomy)—including the first two laws of planetary motion—began with the analysis, under Tycho's direction, of Mars' orbit. Kepler calculated and recalculated various approximations of Mars' orbit using an equant (the mathematical tool that Copernicus had eliminated with his system), eventually creating a model that generally agreed with Tycho's observations to within two arcminutes (the average measurement error). But he was not satisfied with the complex and still slightly inaccurate result; at certain points the model differed from the data by up to eight arcminutes. The wide array of traditional mathematical astronomy methods having failed him, Kepler set about trying to fit an ovoid orbit to the data.

In Kepler's religious view of the cosmos, the Sun (a symbol of God the Father) was the source of motive force in the solar system. As a physical basis, Kepler drew by analogy on William Gilbert's theory of the magnetic soul of the Earth from De Magnete (1600) and on his own work on optics. Kepler supposed that the motive power (or motive species) radiated by the Sun weakens with distance, causing faster or slower motion as planets move closer or farther from it. Perhaps this assumption entailed a mathematical relationship that would restore astronomical order. Based on measurements of the aphelion and perihelion of the Earth and Mars, he created a formula in which a planet's rate of motion is inversely proportional to its distance from the Sun. Verifying this relationship throughout the orbital cycle, however, required very extensive calculation; to simplify this task, by late 1602 Kepler reformulated the proportion in terms of geometry: planets sweep out equal areas in equal times—Kepler's second law of planetary motion.

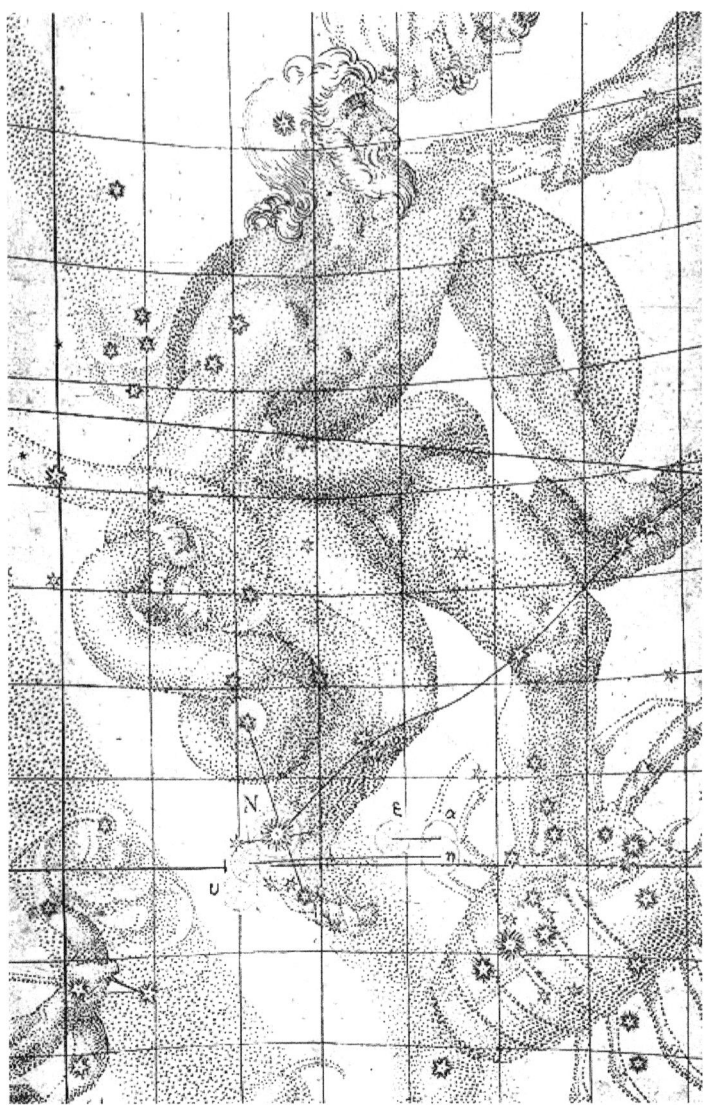

The location of the stella nova, in the foot of Ophiuchus, is marked with an N (8 grid squares down, 4 over from the left).

He then set about calculating the entire orbit of Mars, using the geometrical rate law and assuming an egg-shaped ovoid orbit. After approximately 40 failed attempts, in early 1605 he at last hit upon the idea of an ellipse, which he had previously assumed to be too simple a solution for earlier astronomers to have overlooked. Finding that an elliptical orbit fit the Mars data, he immediately concluded that all planets move in ellipses, with the sun at one focus—Kepler's first law of planetary motion. Because he employed no calculating assistants, however, he did not extend the mathematical analysis beyond Mars. By the end of the year, he completed the manuscript for Astronomia nova, though it would not be published until 1609 due to legal disputes over the use of Tycho's observations, the property of his heirs.

Dioptrice, Somnium Manuscript, and Other Work

In the years following the completion of Astronomia Nova, most of Kepler's research was focused on preparations for the Rudolphine Tables and a comprehensive set of ephemerides (specific predictions of planet and star positions) based on the table (though neither would be completed for many years). He also attempted (unsuccessfully) to begin a collaboration with Italian astronomer Giovanni Antonio Magini. Some of his other work dealt with chronology, especially the dating of events in the life of Jesus, and with astrology, especially criticism of dramatic predictions of catastrophe such as those of Helisaeus Roeslin.

DE MOTIB. STELLÆ MARTIS

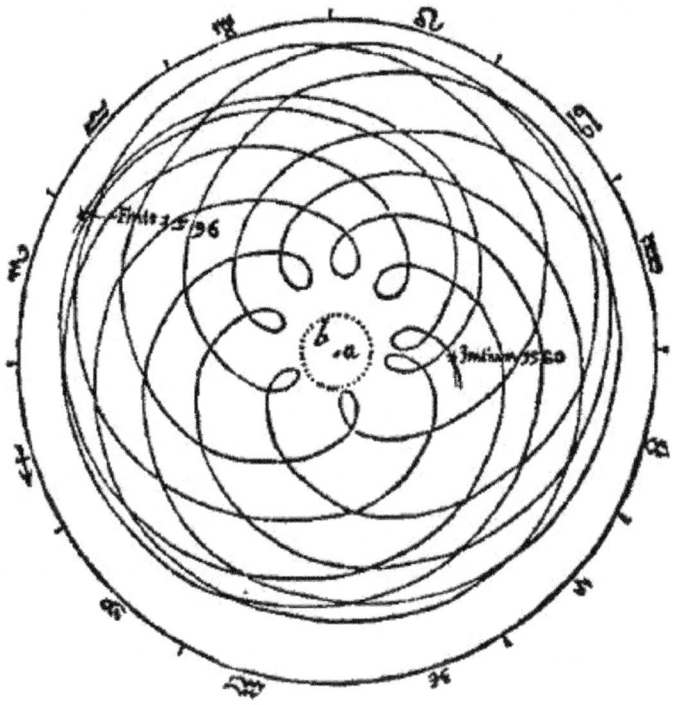

Kepler and Roeslin engaged in a series of published attacks and counter-attacks, while physician Philip Feselius published a work dismissing astrology altogether (and Roeslin's work in particular). In response to what Kepler saw as the excesses of astrology on the one hand and overzealous rejection of it on the other,

Kepler prepared Tertius Interveniens [Third-party Interventions]. Nominally this work—presented to the common patron of Roeslin and Feselius—was a neutral mediation between the feuding scholars, but it also set out Kepler's general views on the value of astrology, including some hypothesized mechanisms of interaction between planets and individual souls. While Kepler considered most traditional rules and methods of astrology to be the "evil-smelling dung" in which "an industrious hen" scrapes, there was an "occasional grain-seed, indeed, even a pearl or a gold nugget" to be found by the conscientious scientific astrologer. Conversely, Sir Oliver Lodge observed that Kepler was somewhat disdainful of astrology, as Kepler was "continually attacking and throwing sarcasm at astrology, but it was the only thing for which people would pay him, and on it after a fashion he lived."

Karlova street in Old Town, Prague – house
where Kepler lived. Museum

- ROBERT STAWELL BALL

In the first months of 1610, Galileo Galilei—using his powerful new telescope—discovered four satellites orbiting Jupiter. Upon publishing his account as Sidereus Nuncius [Starry Messenger], Galileo sought the opinion of Kepler, in part to bolster the credibility of his observations. Kepler responded enthusiastically with a short published reply, Dissertatio cum Nuncio Sidereo [Conversation with the Starry Messenger]. He endorsed Galileo's observations and offered a range of speculations about the meaning and implications of Galileo's discoveries and telescopic methods, for astronomy and optics as well as cosmology and astrology. Later that year, Kepler published his own telescopic observations of the moons in Narratio de Jovis Satellitibus, providing further support of Galileo. To Kepler's disappointment, however, Galileo never published his reactions (if any) to Astronomia Nova.

After hearing of Galileo's telescopic discoveries, Kepler also started a theoretical and experimental investigation of telescopic optics using a telescope borrowed from Duke Ernest of Cologne. The resulting manuscript was completed in September 1610 and published as Dioptrice in 1611. In it, Kepler set out the theoretical basis of double-convex converging lenses and double-concave diverging lenses—and how they are combined to produce a Galilean telescope—as well as the concepts of real vs. virtual images, upright vs. inverted images, and the effects of focal length on magnification and reduction. He also described an improved telescope—now known as the astronomical or Keplerian telescope—in which two convex lenses can produce higher magnification than Galileo's combination of convex and concave lenses.

Around 1611, Kepler circulated a manuscript of what would eventually be published (posthumously) as Somnium [The Dream]. Part of the purpose of Somnium was to describe what practicing astronomy would be like from the perspective of another planet, to show the feasibility of a non-geocentric system. The manuscript, which disappeared after changing hands several times, described a fantastic trip to the moon; it was part allegory, part autobiography, and part treatise on interplanetary travel (and is sometimes described as the first work of science fiction). Years later, a distorted version of the story may have instigated the witchcraft trial against his mother, as the mother of the narrator consults a demon to learn the means of space travel. Following her eventual acquittal, Kepler composed 223 footnotes to the story—several times longer than the actual text—which explained the allegorical aspects as well as the considerable scientific content (particularly regarding lunar geography) hidden within the text.

Work in Mathematics and Physics

As a New Year's gift that year (1611), he also composed for his friend and some-time patron, Baron Wackher von Wackhenfels, a short pamphlet entitled Strena Seu de Nive Sexangula (A New Year's Gift of Hexagonal Snow). In this treatise, he published the first description of the hexagonal symmetry of snowflakes and, extending the discussion into a hypothetical atomistic physical basis for the symmetry, posed what later became known as the Kepler conjecture, a statement about the most efficient arrangement for packing spheres.

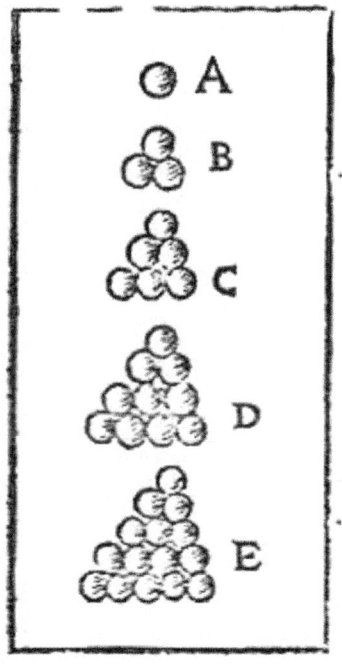

One of the diagrams from Strena Seu de Nive Sexangula, illustrating the Kepler conjecture

Personal and Political Troubles

In 1611, the growing political-religious tension in Prague came to a head. Emperor Rudolph—whose health was failing—was forced to abdicate as King of Bohemia by his brother Matthias. Both sides sought Kepler's astrological advice, an opportunity he used to deliver conciliatory political advice (with little reference to the stars, except in general statements to discourage drastic action). However, it was clear that Kepler's future prospects in the court of Matthias were dim.

Also in that year, Barbara Kepler contracted Hungarian spotted fever, then began having seizures. As Barbara was recovering, Kepler's three children all fell sick with smallpox; Friedrich, 6, died. Following his son's death, Kepler sent letters to potential patrons in Württemberg and Padua. At the University of Tübingen in Württemberg, concerns over Kepler's perceived Calvinist heresies in violation of the Augsburg Confession and the Formula of Concord prevented his return. The University of Padua—on the recommendation of the departing Galileo—sought Kepler to fill the mathematics professorship, but Kepler, preferring to keep his family in German territory, instead travelled to Austria to arrange a position as teacher and district mathematician in Linz. However, Barbara relapsed into illness and died shortly after Kepler's return.

Kepler postponed the move to Linz and remained in Prague until Rudolph's death in early 1612, though between political upheaval, religious tension, and family tragedy (along with the legal dispute over his wife's estate), Kepler could do no research. Instead, he pieced together a chronology manuscript, Eclogae Chronicae, from correspondence and earlier work. Upon succession as Holy Roman Emperor, Matthias re-affirmed Kepler's position (and salary) as imperial mathematician but allowed him to move to Linz.

Linz and Elsewhere (1612–1630)

In Linz, Kepler's primary responsibilities (beyond completing the Rudolphine Tables) were teaching at the district school and providing astrological and astronomical services. In his first years there, he enjoyed financial security and religious freedom relative to his life in Prague—though he was excluded from Eucharist by his Lutheran church over his theological scruples. His first publication in Linz was De vero Anno (1613), an expanded treatise on the year of Christ's birth; he also participated in deliberations on whether to introduce Pope Gregory's reformed calendar to Protestant German lands; that year he also wrote the influential mathematical treatise Nova stereometria doliorum vinariorum, on measuring the volume of containers such as wine barrels, published in 1615.

A statue of Kepler in Linz

Second Marriage

On October 30, 1613, Kepler married the 24-year-old Susanna Reuttinger. Following the death of his first wife Barbara, Kepler had considered 11 different matches over two years (a decision process formalized later as the marriage problem). He eventually returned to Reuttinger (the fifth match) who, he wrote, "won me over with love, humble loyalty, economy of household, diligence, and the love she gave the stepchildren." The first three children of this marriage (Margareta Regina, Katharina, and Sebald) died in childhood. Three more survived into adulthood: Cordula (born 1621); Fridmar (born 1623); and Hildebert (born 1625). According to Kepler's biographers, this was a much happier marriage than his first.

Epitome of Copernican Astronomy, Calendars, and the witch trial of his Mother

Since completing the Astronomia nova, Kepler had intended to compose an astronomy textbook. In 1615, he completed the first of three volumes of Epitome astronomiae Copernicanae (Epitome of Copernican Astronomy); the first volume (books I–III) was printed in 1617, the second (book IV) in 1620, and the third (books V–VII) in 1621. Despite the title, which referred simply to heliocentrism, Kepler's textbook culminated in his own ellipse-based system. The Epitome became Kepler's most influential work. It contained all three laws of planetary motion and attempted to explain heavenly motions through physical causes.[58] Though it explicitly extended the first two laws of planetary motion (applied to Mars in Astronomia nova) to all the planets as well as the Moon and the Medicean satellites of Jupiter, it did not explain how elliptical orbits could be derived from observational data.

As a spin-off from the Rudolphine Tables and the related Ephemerides, Kepler published astrological calendars, which were very popular and helped offset the costs of producing his other work—especially when support from the Imperial treasury was withheld. In his calendars—six between 1617 and 1624—Kepler forecast planetary positions and weather as well as political events; the latter were often cannily accurate, thanks to his keen grasp of contemporary political and theological tensions. By 1624, however, the escalation of those tensions and the ambiguity of the prophecies meant political trouble for Kepler himself; his final calendar was publicly burned in Graz.

- ROBERT STAWELL BALL

Kepler's Figure 'M' from the Epitome,
showing the world as belonging to just
one of any number of similar stars.

In 1615, Ursula Reingold, a woman in a financial dispute with Kepler's brother Christoph, claimed Kepler's mother Katharina had made her sick with an evil brew. The dispute escalated, and in 1617 Katharina was accused of witchcraft; witchcraft trials were relatively common in central Europe at this time. Beginning in August 1620, she was imprisoned for fourteen months. She was released in October 1621, thanks in part to the extensive legal defense drawn up by Kepler. The accusers had no stronger evidence than rumors. Katharina was subjected to territio verbalis, a graphic description of the torture awaiting her as a witch, in a final attempt to make her confess. Throughout the trial, Kepler postponed his other work to focus on his "harmonic theory". The result, published in 1619, was Harmonices Mundi ("Harmony of the World").

Harmonices Mundi

Kepler was convinced "that the geometrical things have provided the

Creator with the model for decorating the whole world". In Harmony, he attempted to explain the proportions of the natural world—particularly the astronomical and astrological aspects—in terms of music. The central set of "harmonies" was the musica universalis or "music of the spheres", which had been studied by Pythagoras, Ptolemy and many others before Kepler; in fact, soon after publishing Harmonices Mundi, Kepler was embroiled in a priority dispute with Robert Fludd, who had recently published his own harmonic theory.

Kepler began by exploring regular polygons and regular solids, including the figures that would come to be known as Kepler's solids. From there, he extended his harmonic analysis to music, meteorology, and astrology; harmony resulted from the tones made by the souls of heavenly bodies—and in the case of astrology, the interaction between those tones and human souls. In the final portion of the work (Book V), Kepler dealt with planetary motions, especially relationships between orbital velocity and orbital distance from the Sun. Similar relationships had been used by other astronomers, but Kepler—with Tycho's data and his own astronomical theories— treated them much more precisely and attached new physical significance to them.

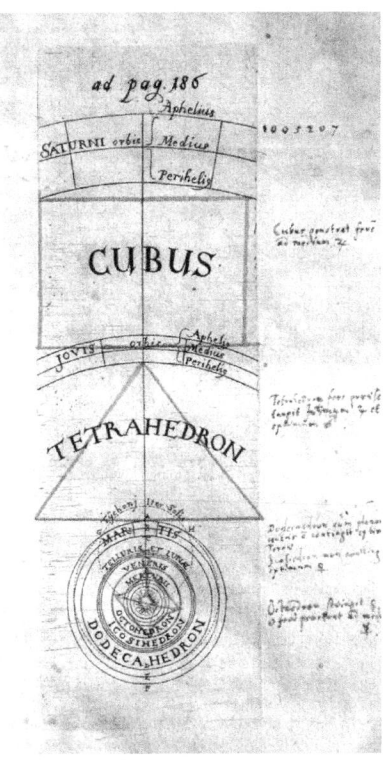

Geometrical harmonies in the perfect solids from Harmonices Mundi (1619)

86

Among many other harmonies, Kepler articulated what came to be known as the third law of planetary motion. He then tried many combinations until he discovered that (approximately) "The square of the periodic times are to each other as the cubes of the mean distances." Although he gives the date of this epiphany (March 8, 1618), he does not give any details about how he arrived at this conclusion. However, the wider significance for planetary dynamics of this purely kinematical law was not realized until the 1660s. When conjoined with Christiaan Huygens' newly discovered law of centrifugal force, it enabled Isaac Newton, Edmund Halley, and perhaps Christopher Wren and Robert Hooke to demonstrate independently that the presumed gravitational attraction between the Sun and its planets decreased with the square of the distance between them. This refuted the traditional assumption of scholastic physics that the power of gravitational attraction remained constant with distance whenever it applied between two bodies, such as was assumed by Kepler and also by Galileo in his mistaken universal law that gravitational fall is uniformly accelerated, and also by Galileo's student Borrelli in his 1666 celestial mechanics.

Rudolphine Tables and his last years

In 1623, Kepler at last completed the Rudolphine Tables, which at the time was considered his major work. However, due to the publishing requirements of the emperor and negotiations with Tycho Brahe's heir, it would not be printed until 1627. In the meantime, religious tension — the root of the ongoing Thirty Years' War — once again put Kepler and his family in jeopardy. In 1625, agents of the Catholic Counter-Reformation placed most of Kepler's library under seal, and in 1626 the city of Linz was besieged. Kepler moved to Ulm, where he arranged for the printing of the Tables at his own expense.

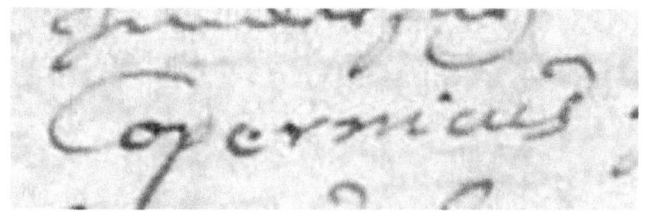

Name "Copernicus" in a manuscript report by
Kepler concerning the Rudolphine Tables (1616).

In 1628, following the military successes of
the Emperor Ferdinand's armies under
General Wallenstein, Kepler became an
official advisor to Wallenstein. Though not
the general's court astrologer per se,
Kepler provided astronomical calculations
for Wallenstein's astrologers and
occasionally wrote horoscopes himself. In
his final years, Kepler spent much of his
time traveling, from the imperial court in
Prague to Linz and Ulm to a temporary
home in Sagan, and finally to Regensburg.
Soon after arriving in Regensburg, Kepler
fell ill. He died on November 15, 1630, and
was buried there; his burial site was lost
after the Swedish army destroyed the
churchyard. Only Kepler's self-authored
poetic epitaph survived the times:

Mensus eram coelos, nunc terrae metior umbras
Mens coelestis erat, corporis umbra iacet.

I measured the skies, now the shadows I measure
Skybound was the mind, earthbound the body rests.

Reception of his Astronomy

Kepler's laws were not immediately accepted. Several major figures such as Galileo and René Descartes completely ignored Kepler's Astronomia nova. Many astronomers, including Kepler's teacher, Michael Maestlin, objected to Kepler's introduction of physics into his astronomy. Some adopted compromise positions. Ismael Boulliau accepted elliptical orbits but replaced Kepler's area law with uniform motion in respect to the empty focus of the ellipse, while Seth Ward used an elliptical orbit with motions defined by an equant.

T A B U L Æ

RUDOLPHINÆ,

QVIBVS ASTRONOMICÆ SCIENTIÆ, TEMPO-
rum longinquitate collapfæ RESTAURATIO continetur;

A Phœnice illo Aftronomorum

TYCHONE

Ex Illuftri & Generofa BRAHEORUM in Regno Daniæ
familiâ oriundo Equite,

PRIMUM ANIMO CONCEPTA ET DESTINATA ANNO
CHRISTI MDLXIV: EXINDE OBSERVATIONIBUS SIDERUM ACCURA-
TISSIMIS, POST ANNUM PRÆCIPUE MDLXXII, QUO SIDUS IN CASSIOPEJÆ
CONSTELLATIONE NOVUM EFFULSIT, SERIÒ AFFECTATA; VARIISQUE OPERIBUS, CÙM ME-
chanicis, tùm librariis, impenfo patrimonio ampliffimo, accedentibus etiam fubfidiis FRIDERICI II. DANIÆ
REGIS, regali magnificentia dignis, tractâ per annos XXV, potiffimùm in Infula freti SUNDICI HUEN-
NA, & arce URANIBURGO, in hos ufus à fundamentis extructâ:

TANDEM TRADUCTA IN GERMANIAM, INQVE AVLAM ET
Nomen RUDOLPHI IMP. anno MDIIC.

TABULAS IPSAS, JAM ET NUNCUPATAS, ET AFFECTAS, SED
MORTE AUTHORIS SUI ANNO MDCI DESERTAS,

JVSSV ET STIPENDIIS FRETVS TRIVM IMPPP.

RUDOLPHI, MATTHIÆ, FERDINANDI,

ANNITENTIBUS HÆREDIBUS BRAHEANIS; EX FUNDAMENTIS
obfervationum relictarum; ad exemplum ferè partium jam exftructarum; continuî multorum annorum fpe-
culationibus, & computationibus, primùm PRAGÆ Bohemorum continuavit; deindè LINCII,
fuperioris Auftriæ Metropoli, fubfidiis etiam Ill. Provincialium adjutus, emendavit, per-
fecit, abfolvit : adq, caufarum & calculi perennis formulam traduxit

IOANNES KEPLERUS,

TICHONI primùm à RUDOLPHO II. Imp. adjunctus calculi minifter; indéq,
trium ordine Imppp. Mathematicus:

Qui idem de fpeciali mandato FERDINANDI II. IMP.
petentibus inftantibúsq; Hæredibus,

Opus hoc ad ufus præfentium & pofteritatis, typis, numericis propriis, cæteris & prælo
JONÆ SAURII, Reip. Ulmanæ Typographi, in publicum extulit, &
Typographicis operis ULMÆ curator affuit.

Cum Privilegiis, IMP. & Regum Rerúmq; publ. vivo TYCHONI ejúsq; Hæredibus,
& fpeciali Imperatorio, ipfi KEPLERO conceffo, ad annos XXX.

ANNO M. DC. XXVII.

Title page of the Tabulae Rudolphinae, Ulm, 1627

JOHANNES KEPLER - GREAT ASTRONOMERS

Several astronomers tested Kepler's theory, and its various modifications, against astronomical observations. Two transits of Venus and Mercury across the face of the sun provided sensitive tests of the theory, under circumstances when these planets could not normally be observed. In the case of the transit of Mercury in 1631, Kepler had been extremely uncertain of the parameters for Mercury, and advised observers to look for the transit the day before and after the predicted date. Pierre Gassendi observed the transit on the date predicted, a confirmation of Kepler's prediction. This was the first observation of a transit of Mercury. However, his attempt to observe the transit of Venus just one month later was unsuccessful due to inaccuracies in the Rudolphine Tables. Gassendi did not realize that it was not visible from most of Europe, including Paris. Jeremiah Horrocks, who observed the 1639 Venus transit, had used his own observations to adjust the parameters of the Keplerian model, predicted the transit, and then built apparatus to observe the transit.

- ROBERT STAWELL BALL

He remained a firm advocate of the Keplerian model.

Epitome of Copernican Astronomy was read by astronomers throughout Europe, and following Kepler's death it was the main vehicle for spreading Kepler's ideas. Between 1630 and 1650, it was the most widely used astronomy textbook, winning many converts to ellipse-based astronomy. However, few adopted his ideas on the physical basis for celestial motions. In the late 17th century, a number of physical astronomy theories drawing from Kepler's work—notably those of Giovanni Alfonso Borelli and Robert Hooke—began to incorporate attractive forces (though not the quasi-spiritual motive species postulated by Kepler) and the Cartesian concept of inertia. This culminated in Isaac Newton's Principia Mathematica (1687), in which Newton derived Kepler's laws of planetary motion from a force-based theory of universal gravitation.

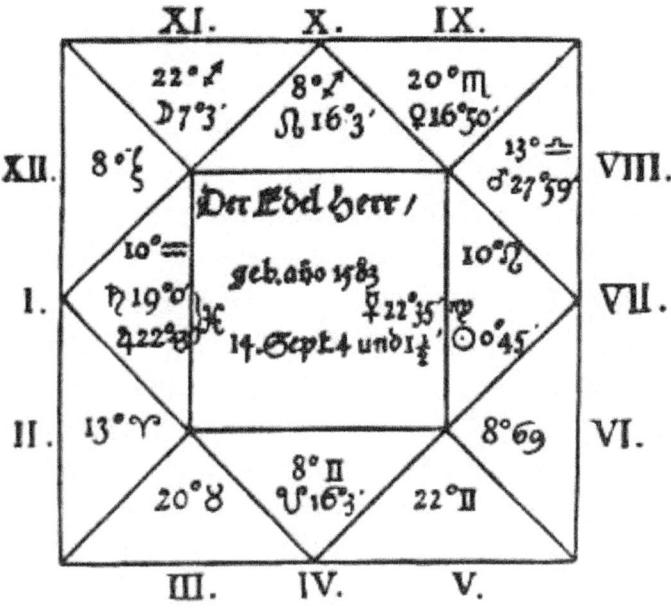

Kepler's horoscope for General Wallenstein

Historical and Cultural Legacy

History of Science

Beyond his role in the historical development of astronomy and natural philosophy, Kepler has loomed large in the philosophy and historiography of science. Kepler and his laws of motion were central to early histories of astronomy such as Jean-Étienne Montucla's 1758 Histoire des mathématiques and Jean-Baptiste Delambre's 1821 Histoire de l'astronomie moderne. These and other histories written from an Enlightenment perspective treated Kepler's metaphysical and religious arguments with skepticism and disapproval, but later Romantic-era natural philosophers viewed these elements as central to his success. William Whewell, in his influential History of the Inductive Sciences of 1837, found Kepler to be the archetype of the inductive scientific genius; in his Philosophy of the Inductive Sciences of 1840, Whewell held Kepler up as the embodiment of the most advanced forms of scientific method. Similarly, Ernst Friedrich Apelt—the first to extensively study Kepler's manuscripts, after their purchase by Catherine the Great—identified Kepler as a key to the "Revolution of the sciences". Apelt, who saw Kepler's mathematics, aesthetic sensibility, physical ideas, and theology as part of a unified system of thought, produced the first extended analysis of Kepler's life and work.

Alexandre Koyré's work on Kepler was, after Apelt, the first major milestone in historical interpretations of Kepler's cosmology and its influence. In the 1930s and 1940s, Koyré, and a number of others in the first generation of professional historians of science, described the "Scientific Revolution" as the central event in the history of science, and Kepler as a (perhaps the) central figure in the revolution. Koyré placed Kepler's theorization, rather than his empirical work, at the center of the intellectual transformation from ancient to modern world-views. Since the 1960s, the volume of historical Kepler scholarship has expanded greatly, including studies of his astrology and meteorology, his geometrical methods, the role of his religious views in his work, his literary and rhetorical methods, his interaction with the broader cultural and philosophical currents of his time, and even his role as an historian of science.

Philosophers of science—such as Charles Sanders Peirce, Norwood Russell Hanson, Stephen Toulmin, and Karl Popper—have repeatedly turned to Kepler: examples of incommensurability, analogical reasoning, falsification, and many other philosophical concepts have been found in Kepler's work. Physicist Wolfgang Pauli even used Kepler's priority dispute with Robert Fludd to explore the implications of analytical psychology on scientific investigation.

Monument to **Tycho Brahe** and **Kepler**
in Prague, Czech Republic.

- ROBERT STAWELL BALL

Editions and Translations

Modern translations of a number of Kepler's books appeared in the late-nineteenth and early-twentieth centuries, the systematic publication of his collected works began in 1937 (and is nearing completion in the early 21st century).

An edition in eight volumes, Kepleri Opera omnia, was prepared by Christian Frisch (1807–1881), during 1858 to 1871, on the occasion of Kepler's 300th birthday. Frisch's edition only included Kepler's Latin, with a Latin commentary.

A new edition was planned beginning in 1914 by Walther von Dyck (1856–1934). Dyck compiled copies of Kepler's unedited manuscripts, using international diplomatic contacts to convince the Soviet authorities to lend him the manuscripts kept in Leningrad for photographic reproduction. These manuscripts contained several works by Kepler that had not been available to Frisch. Dyck's photographs remain the basis for the modern editions of Kepler's unpublished manuscripts.

The GDR stamp featuring Kepler

Max Caspar (1880–1956) published his German translation of Kepler's Mysterium Cosmographicum in 1923. Both Dyck and Caspar were influenced in their interest in Kepler by mathematician Alexander von Brill (1842–1935). Caspar became Dyck's collaborator, succeeding him as project leader in 1934, establishing the Kepler-Kommission in the following year. Assisted by Martha List (1908–1992) and Franz Hammer (1898–1979), Caspar continued editorial work during World War II. Max Caspar also published a biography of Kepler in 1948. The commission was later chaired by Volker Bialas (during 1976–2003) and Ulrich Grigull (during 1984–1999) and Roland Bulirsch (1998–2014).

Popular science and Historical Fiction

Kepler has acquired a popular image as an icon of scientific modernity and a man before his time; science popularizer Carl Sagan described him as "the first astrophysicist and the last scientific astrologer".

The debate over Kepler's place in the Scientific Revolution has produced a wide variety of philosophical and popular treatments. One of the most influential is Arthur Koestler's 1959 The Sleepwalkers, in which Kepler is unambiguously the hero (morally and theologically as well as intellectually) of the revolution.

A well-received, if fanciful, historical novel by John Banville, Kepler (1981), explored many of the themes developed in Koestler's non-fiction narrative and in the philosophy of science. Somewhat more fanciful is a recent work of nonfiction, Heavenly Intrigue (2004), suggesting that Kepler murdered Tycho Brahe to gain access to his data.

Veneration and eponymy

In Austria, Kepler left behind such a historical legacy that he was one of the motifs of a silver collector's coin: the 10-euro Johannes Kepler silver coin, minted on September 10, 2002. The reverse side of the coin has a portrait of Kepler, who spent some time teaching in Graz and the surrounding areas. Kepler was acquainted with Prince Hans Ulrich von Eggenberg personally, and he probably influenced the construction of Eggenberg Castle (the motif of the obverse of the coin). In front of him on the coin is the model of nested spheres and polyhedra from Mysterium Cosmographicum.

The German composer Paul Hindemith wrote an opera about Kepler entitled Die Harmonie der Welt, and a symphony of the same name was derived from music for the opera. Philip Glass wrote an opera called Kepler based on Kepler's life (2009).

Kepler is honored together with Nicolaus Copernicus with a feast day on the liturgical calendar of the Episcopal Church (USA) on May 23.

Directly named for Kepler's contribution to science are Kepler's laws of planetary motion, Kepler's Supernova (Supernova 1604, which he observed and described) and the Kepler Solids, a set of geometrical constructions, two of which were described by him, and the Kepler conjecture on sphere packing.

The Kepler crater as photographed
by Apollo 12 in 1969

- In astronomy: The lunar crater Kepler (Keplerus, named by Giovanni Riccioli, 1651), the asteroid 1134 Kepler (1929), Kepler (crater on Mars) (1973), Kepler Launch Site for model rockets (2001), the Kepler Mission, a space photometer launched by NASA in 2009, Johannes Kepler ATV (Automated Transfer Vehicle launched to resupply the ISS in 2011).

- Educational institutions: Johannes Kepler University of Linz (1975), Kepler College (Seattle, Washington), besides several institutions of primary and secondary education, such as Johannes Kepler Grammar School, at the site where Kepler lived in Prague, and Kepler Gymnasium, Tübingen

- Streets or squares named after him: Keplerplatz Vienna (station of Vienna U-Bahn), Keplerstraße in Hanau near Frankfurt am Main, Keplerstraße in Munich, Germany,

Keplerstraße and Keplerbrücke in Graz, Austria, Keplerova ulice in Prague.

- The Kepler Mountains and Kepler Track in Fiordland National Park, South Island, New Zealand; Kepler Challenge (1988).

- Kepler, a high end graphics processing microarchitecture introduced by Nvidia in 2012.

Works

- ❖ Mysterium Cosmographicum (The Sacred Mystery of the Cosmos) (1596)

- ❖ De Fundamentis Astrologiae Certioribus (On Firmer Fundaments of Astrology; 1601)
- ❖ Astronomiae Pars Optica (The Optical Part of Astronomy) (1604)

- ❖ De Stella nova in pede Serpentarii (On the New Star in Ophiuchus's Foot) (1606)

- ❖ Astronomia nova (New Astronomy) (1609)

- ❖ Tertius Interveniens (Third-party Interventions) (1610)

- ❖ Dissertatio cum Nuncio Sidereo (Conversation with the Starry Messenger) (1610)

- ❖ Dioptrice (1611)

- ❖ De nive sexangula (On the Six-Cornered Snowflake) (1611)

- ❖ De vero Anno, quo aeternus Dei Filius humanam naturam in Utero benedictae Virginis Mariae assumpsit (1614)

- ❖ Eclogae Chronicae (1615, published with Dissertatio cum Nuncio Sidereo)

- ❖ Nova stereometria doliorum vinariorum (New Stereometry of Wine Barrels) (1615)

- ❖ Epitome astronomiae Copernicanae (Epitome of Copernican Astronomy) (published in three parts from 1618 to 1621)

- ❖ Harmonices Mundi (Harmony of the Worlds) (1619)

- ❖ Mysterium cosmographicum (The Sacred Mystery of the Cosmos), 2nd edition (1621)

❖ Tabulae Rudolphinae (Rudolphine Tables) (1627)

❖ Somnium (The Dream) (1634)

ABOUT THE AUTHOR

ROBERT STAWELL BALL
(July 1, 1840 – November 25, 1913)

Sir Robert Stawell Ball was an Irish astronomer who founded the screw theory.

He was the son of naturalist Robert Ball and Amelia Gresley Hellicar. He was born in Dublin.

- ROBERT STAWELL BALL

Ball worked for Lord Rosse from 1865 to 1867. In 1867 he became Professor of Applied Mathematics at the Royal College of Science in Dublin. There he lectured on mechanics and published an elementary account of the science.

In 1874 Ball was appointed Royal Astronomer of Ireland and Andrews Professor of Astronomy in the University of Dublin at Dunsink Observatory.

Ball contributed to the science of kinematics by delineating the screw displacement:

When Ball and the screw theorists speak of screws they no longer mean actual cylindrical objects with helical threads cut into them but the possible motion of any body whatsoever, including that of the screw independently of the nut.

Ball's treatise *The Theory of Screws* (1876) is now in the public domain. His work on screw dynamics earned him in 1879 the

Cunningham Medal of the Royal Irish Academy.

In 1882 Popular Science Monthly carried his article "*A Glimpse through the Corridors of Time*". The following year it carried his two-part article on "*The Boundaries of Astronomy*".

Ball expounded the tides in *Time and Tide*: a Romance of the Moon In 1892 he was appointed Lowndean Professor of Astronomy and Geometry at Cambridge University at the same time becoming director of the Cambridge Observatory. He was a fellow of King's College, Cambridge.

In 1900 Cambridge University Press published *A Treatise on the Theory of Screws*. That year he also published *The Story of the Heavens*. Much in the limelight, he stood as President of the Quaternion Society. He was also President of the Mathematical Association in 1900.

In 1908 he published *A Treatise on Spherical Astronomy*, which is a textbook on astronomy starting from spherical trigonometry and the celestial sphere, considering atmospheric refraction and aberration of light, and introducing basic use of a generalised instrument.

His work *The Story of the Heavens* is mentioned in the "Ithaka" chapter of Ulysses. His lectures, articles and books (e.g. *Starland* and *The Story of the Heavens*) were mostly popular and simple in style.

He died in Cambridge and was buried at the Parish of the Ascension Burial Ground in Cambridge, with his wife Lady Francis Elizabeth Ball. Their children were: Frances Amelia, Robert Steele, William Valentine (later Sir), Mary Agnetta, Charles Rowan Hamilton, and Randall Gresley (later Colonel).

LECTURES

Ball became celebrated for his popular lectures on science. He gave an estimated 2500 lectures between 1875 and 1910 in towns and cities across Britain and Ireland. In 1892, 1898 and 1900 he was invited to deliver the Royal Institution Christmas Lecture. *Astronomy; Astronomy* and *Great Chapters from the Book of Nature.*

FEEDBACK

Now that you have
read the book ...

Was it interesting?

Did you enjoy what you wanted to read?
Was there any room for improvement?

Let us know at:
http://www.diamondbooks.ca/feedback

Your feedback is highly appreciated.
Thank you!

- ROBERT STAWELL BALL

Would you like to buy a copy of

'JOHANNES KEPLER'
by Robert Stawell Ball?

Order Online!

PLEASE VISIT:
http://www.diamondbooks.ca

DIAMOND™
BOOKS
www.diamondbooks.ca

HUGE SAVINGS ON BULK ORDERS
(10 copies, 20 copies, 50 copies, 100 copies, 500 copies, 1000 copies)

Please send your request at:
http://www.diamondbooks.ca/bulkorder

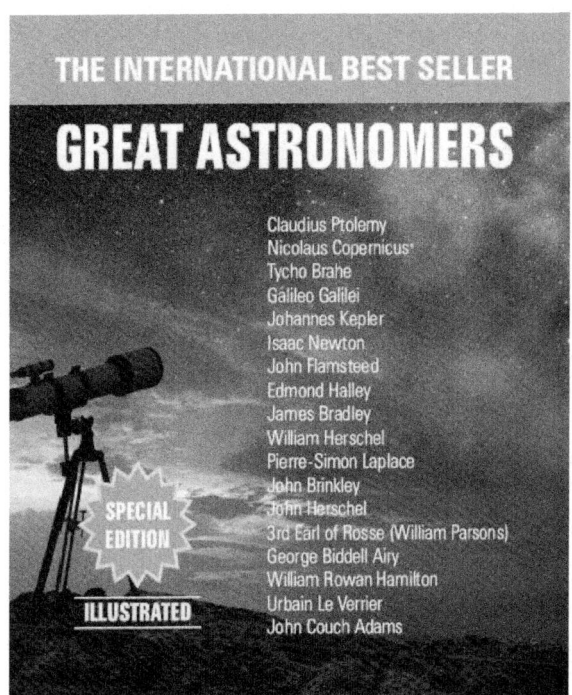

Would you like to buy a copy of
'GREAT ASTRONOMERS'

PLEASE VISIT:
http://www.diamondbooks.ca

Would you like to buy a copy of 'NICOLAUS COPERNICUS?'

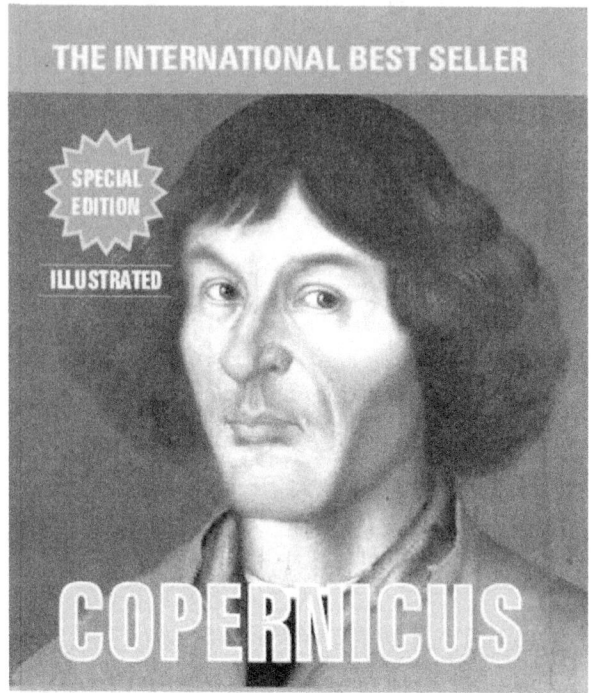

Would you like to buy a copy of 'JOHANNES KEPLER?'

PLEASE VISIT:

http://www.diamondbooks.ca

Would you like to buy a copy of 'ISSAC NEWTON?'

THE INTERNATIONAL BEST SELLER

SPECIAL EDITION

ILLUSTRATED

NEWTON

Isaac Newton (December 25, 1642 – March 20, 1726/27)
AN ENGLISH PHYSICIST AND MATHEMATICIAN

ROBERT STAWELL BALL

NOTES

NOTES

NOTES

NOTES

NOTES

NOTES

Printed in Great Britain
by Amazon

31879746R00086